Army Robotics and Artificial Intelligence

A 1987 Review

Committee to Review Army Robotics and Artificial Intelligence
Manufacturing Studies Board
Commission on Engineering and Technical Systems
National Research Council

NATIONAL ACADEMY PRESS
Washington, D.C.   1987

NOTICE: The project that is the subject of this report was approved by the Governing Board of the National Research Council, whose members are drawn from the councils of the National Academy of Sciences, the National Academy of Engineering, and the Institute of Medicine. The members of the committee responsible for the report were chosen for their special competences and with regard for appropriate balance.

This report has been reviewed by a group other than the authors according to procedures approved by a Report Review Committee consisting of members of the National Academy of Sciences, the National Academy of Engineering, and the Institute of Medicine.

The National Academy of Sciences is a private, nonprofit, self-perpetuating society of distinguished scholars engaged in scientific and engineering research, dedicated to the furtherance of science and technology and to their use for the general welfare. Upon the authority of the charter granted to it by the Congress in 1863, the Academy has a mandate that requires it to advise the federal government on scientific and technical matters. Dr. Frank Press is president of the National Academy of Sciences.

The National Academy of Engineering was established in 1964, under the charter of the National Academy of Sciences, as a parallel organization of outstanding engineers. It is autonomous in its administration and in the selection of its members, sharing with the National Academy of Sciences the responsibility for advising the federal government. The National Academy of Engineering also sponsors engineering programs aimed at meeting national needs, encourages education and research, and recognizes the superior achievements of engineers. Dr. Robert M. White is president of the National Academy of Engineering.

The Institute of Medicine was established in 1970 by the National Academy of Sciences to secure the services of eminent members of appropriate professions in the examination of policy matters pertaining to the health of the public. The Institute acts under the responsibility given to the National Academy of Sciences by its congressional charter to be an adviser to the federal government and, upon its own initiative, to identify issues of medical care, research, and education. Dr. Samuel O. Thier is president of the Institute of Medicine.

The National Research Council was organized by the National Academy of Sciences in 1916 to associate the broad community of science and technology with the Academy's purposes of furthering knowledge and advising the federal government. Functioning in accordance with general policies determined by the Academy, the Council has become the principal operating agency of both the National Academy of Sciences and the National Academy of Engineering in providing services to the government, the public, and the scientific and engineering communities. The Council is administered jointly by both Academies and the Institute of Medicine. Dr. Frank Press and Dr. Robert M. White are chairman and vice chairman, respectively, of the National Research Council.

This study was supported by Contract No. DACA72-85-C-0006 between the United States Army and the National Academy of Sciences.

A limited number of copies
are available from:

Manufacturing Studies Board
National Research Council
2101 Constitution Avenue
Washington, D.C. 20418

Printed in the United States of America

COMMITTEE TO REVIEW ARMY ROBOTICS AND ARTIFICIAL INTELLIGENCE

WALTER L. ABEL, <u>Chairman</u>, Vice President (retired), Emhart Corporation, Avon, Connecticut
MARGARET A. EASTWOOD, Director, Integrated Factory Controls, CIMCORP, Inc., Aurora, Illinois
FREDERICK W. FOX, Vice President, Operations, PAX International, Indianapolis, Indiana
LESTER A. GERHARDT, ECSE Department, Rensselaer Polytechnic Institute, Troy, New York
JOHN R. GUTHRIE, General (retired), U.S. Army, Annandale, Virginia
TENHO H. HUKKALA, Senior Analyst, National Security Research Group, System Planning Corporation, Arlington, Virginia
ROGER N. NAGEL, Director, Manufacturing Systems Engineering, Lehigh University, Bethlehem, Pennsylvania
CHARLES A. ROSEN, Chief Scientist, Machine Intelligence Corporation, Atherton, California

STAFF

GEORGE H. KUPER, Executive Director, Manufacturing Studies Board
JANICE E. GREENE, Staff Officer
DENNIS A. DRISCOLL, Staff Associate
LUCY V. FUSCO, Administrative Assistant

MANUFACTURING STUDIES BOARD

WICKHAM SKINNER, Chairman, James E. Robison Professor of Business Administration (emeritus), Harvard University, Boston, Massachusetts

ANDERSON ASHBURN, Editor, AMERICAN MACHINIST, New York, New York

AVAK AVAKIAN, Vice President, GTE Sylvania Systems Group, Waltham, Massachusetts

IRVING BLUESTONE, Professor of Labor Studies, Wayne State University, Detroit, Michigan

BARBARA A. BURNS, Manager, SYSTECON, Division of Coopers & Lybrand, Duluth, Georgia

CHARLES E. EBERLE, Vice President, Engineering (retired), The Procter and Gamble Company, Cincinnati, Ohio

ELLIOTT M. ESTES, President (retired), General Motors Corporation, Detroit, Michigan

ROBERT S. KAPLAN, Arthur Lowes Dickinson Professor of Accounting, Graduate School of Business Administration, Harvard University, Boston, Massachusetts

ROBERT B. KURTZ, Vice President (retired), General Electric Corporation, Fairfield, Connecticut

JAMES F. LARDNER, Vice President, Component Group, Deere & Company, Moline, Illinois

MARTIN J. McHALE, Vice President, Control Data Corporation, Bloomington, Minnesota

THOMAS J. MURRIN, President, Energy and Advanced Technology Group, Westinghouse Electric Company, Pittsburgh, Pennsylvania

ROGER N. NAGEL, Director, Manufacturing Systems Engineering, Lehigh University, Bethlehem, Pennsylvania

RICHARD R. NELSON, H. C. Luce Professor of International Political Economy, Columbia University, New York, New York

DAN L. SHUNK, Director, Center for Automated Engineering and Robotics, Arizona State University, Tempe, Arizona
JEROME A. SMITH, Director of Operations, Martin Marietta Corporation, Bethesda, Maryland
JOHN M. STEWART, Director, McKinsey and Company, Inc., New York, New York
STEVEN C. WHEELWRIGHT, Kleiner Perkins Caulfield & Byers Professor of Management, Stanford University, Stanford, California
JOHN A. WHITE, Regents' Professor of Industrial and Systems Engineering, Georgia Institute of Technology, Atlanta, Georgia
EDWIN M. ZIMMERMAN, Member, D.C. Bar, Washington, D.C.

STAFF

GEORGE H. KUPER, Executive Director
KERSTIN B. POLLACK, Director, Program Development
JANICE E. GREENE, Staff Officer
THOMAS C. MAHONEY, Staff Officer
VERNA J. BOWEN, Administrative Assistant
LUCY V. FUSCO, Administrative Assistant
MICHAEL S. RESNICK, Administrative Assistant

ACKNOWLEDGMENTS

The Committee to Review Army Robotics and Artificial Intelligence is responsible for organizing and conducting the research and writing the findings of this study. Our work would not have been possible, however, without the invaluable contributions of the Manufacturing Studies Board staff who facilitated our work: executive director George Kuper, staff officer Janice Greene, staff associate Dennis Driscoll, and administrative assistant Lucy Fusco.

We also wish to thank the peer reviewers--Philip H. Francis, Ira Jacobson, Robert B. Kelley, Jerome A. Smith, and Arthur R. Thomson. Their thoughtful comments on our draft report enabled us to fine-tune its substance and presentation.

Perhaps most importantly, we wish to thank the many people from the U.S. Army who so generously gave their time to meet with us and whose candor made this report possible. These people were:

Ray E. Bowles, Chief, Mobility Branch, Laboratory Command

Thomas Broach, Office of the Assistant Director for Army Research and Technology

Philip Emmerman, Chief, Advanced Sensor Systems Branch, Harry Diamond Laboratories, Laboratory Command

Larry Gambino, Director, Research Institute, U.S. Army Engineer Topographic Laboratories

Ronald Green, U.S. Army Research Office, Electronics Division

Lucy Hagan, Physical Science Administrator, U.S. Army Materiel Command

Catherine Knudson, Research Psychologist/Staff Officer, U.S. Army Medical Research and Development Command

Robert Leighty, Director (retired), Research Institute, U.S. Army Engineer Topographic Laboratory
Joseph Psotka, U.S. Army Research Institute for Behavioral and Social Sciences
Kenneth Rose, U.S. Army Training and Doctrine Command
Charles Shoemaker, Leader, Robotics Sciences and Military Applications Team, U.S. Army Human Engineering Laboratory
Alex Stewart, Electronics Engineer, Technology Planning & Management Directorate, Laboratory Command
Richard Vitali, Technical Director, U.S. Army Laboratory Command
Harry Wiggins, Deputy Chief of Staff for Operations, U.S. Army Laboratory Command
Bruce Zimmerman, Office of the Deputy Chief of Staff for Research, Development, and Acquisition

                                          Walter L. Abel
                                          Chairman

CONTENTS

1. HISTORY AND SCOPE OF THIS PROJECT. 1

   The Original Committee's 1983 Report, 1
   Activities of this Committee, 3

2. ASSESSMENT OF INDIVIDUAL ARMY PROGRAMS 5

   The Teleoperated Mobile Anti-Armor Program, 6
   Robotic Material Handling Equipment, 7
   Robotic Combat Vehicles, 8
   Hawk Maintenance Tutor, 10
   Legged Machines, 10
   Summary of Technical Areas, 11

3. THE ARMY ENVIRONMENT FOR ROBOTICS AND
   ARTIFICIAL INTELLIGENCE. 15

   Inter- and Intra-Agency Coordination, 15
   The Need for Leadership and a Champion, 17
   Funding of Applications, 17
   Industrial Applications, 18

4. EDUCATION AND TRAINING 20

   The Urgent Army Need, 20
   University Centers Sponsored by the Army, 21
   Army Internal Education Programs, 22
   Army Internal Training Programs, 23

5. CONCLUSIONS AND RECOMMENDATIONS. 24

   Conclusions, 24
   Recommendations, 26

LIST OF ACRONYMS 28

1 HISTORY AND SCOPE OF THIS PROJECT

THE ORIGINAL COMMITTEE'S 1983 REPORT

In 1982 and 1983, a 16-member committee of the National Research Council's Manufacturing Studies Board

1. reviewed U.S. Army activities in robotics and artificial intelligence, and
2. described expected developments in those fields over the next 5- and 10-year periods.

The result of that committee's study was the report, <u>Applications of Robotics to Reduce Risk and Improve Effectiveness</u>, released in October 1983.

In its report, the committee stressed the significant contribution that artificial intelligence (AI) and robotics potentially can make to Army operations. The report also established criteria for the selection of AI and robotics projects to fund, and suggested three primary and three secondary projects based on those criteria. The committee noted that

". . . these technologies can enable the Army to

- improve combat capabilities,
- minimize exposure of personnel to hazardous environments,
- increase mission flexibility,
- increase system reliability,
- reduce unit/life cycle costs,
- reduce manpower requirements,
- simplify training."

For the Army to realize these benefits, the committee recommended the development of some short-term demonstrators that could be progressively upgraded. The committee offered the following criteria for selecting specific demonstrators:

- the project should meet clear Army needs,
- a demonstration should be possible within 2 to 3 years,
- the project should use the best state-of-the-art technology available, and
- computer capacity should be sufficient for future upgrades.

An additional consideration was that the selected projects could form a base for acquainting Army personnel at all levels with these new and revolutionary technologies. As upgraded, the applications would need to be capable of operating in a hostile environment.

The selection of specific applications of these technologies was appropriately left to the Army. The committee did, however, suggest that the following applications met the criteria:

"- The <u>Automatic Loader of Ammunition in Tanks</u>, using a robotic arm to replace the human loader of ammunition in a tank. We recommend that two contractors work simultaneously for 2 to 2-1/2 years at a total cost of $4 to $5 million per contractor.
- The <u>Surveillance/Sentry Robot</u>, a portable, possibly mobile platform to detect and identify movement of troops. Funded at $5 million for 2 to 3 years, the robot should be able to include two or more sensor modalities.
- The <u>Intelligent Maintenance, Diagnosis, and Repair System</u>, in its initial form ($1 million over 2 years), will be an interactive trainer. Within 3 years, for an additional $5 million, the system should be expanded to diagnose and suggest repairs for common breakdowns, recommend whether or not to repair, and record the repair history of a piece of equipment.

If additional funds are available, . . . the medical expert system, the flexible material-handling modules, and the battalion information system are also well worth doing."

ACTIVITIES OF THIS COMMITTEE

At the request of the Office of the Deputy Chief of Staff (Research, Development, and Acquisition), eight of the original committee members convened in 1986 to review Army activities in developing and implementing robotics and artificial intelligence. The committee considered not only responses to its 1983 report, but also new issues in the Army's programs and technical advances since 1983.

The new committee met twice in calendar year 1986 and visited four sites where members viewed planned or potential Army applications in action. An update was given to the committee in early 1987.

The committee believes that the recommendations of its earlier report are not only still valid, but even more applicable today. The declining percentage of 19- to 21-year-olds in the population, and the nation's continuing reliance on technologically superior weapons rather than numerically superior weapons, underscore both the continuing need and the opportunity for the Army to apply robotics and artificial intelligence to its operations. Because those technologies have advanced since the committee wrote its original report, the Army can draw on even more industry and university experience.

During the nearly 4 years that have passed, Army activity in these technologies has increased in quantity and visibility. While the Army's budget for robotics ($25 million in FY 1987) and AI ($20 million in FY 1987) is still much less than its need, Army activities in this area have acquired momentum. The committee found that at present, the Army's two greatest needs in implementing AI and robotics are for increased education at all levels and for high-level leadership to coordinate and support these programs. Nonetheless, the Army's progress is reflected not only in the specific projects but also in a more general awareness of the technologies and issues concerning their use.

Broader issues of program management--including decisions about how long to develop technologies in parallel, internal vs. external development, and when to build a prototype--were beyond the scope of this committee. As the Army's AI and robotics programs continue to grow, such issues will also grow in importance.

This report summarizes the committee's assessment of the Army's recent progress in AI and robotics and makes recommendations that will help the Army exploit these important technologies. Progress on specific 1983 recommendations is shown in Table 1.

TABLE 1 Summary of Army Progress in Robotics and AI, 1983-1987

1983 RECOMMENDATION	1987 STATUS
1. Start using available technology now	1. Some progress, but short-term applications are still underfunded
2. Start with a few short-term applications that are likely to succeed	2. The focus on 3 robotic projects is responsive to this recommendation and highly commendable
3. Plan for long-term upgrades	3. Results are mixed; the Advanced Ground Vehicle Technology has a good long-term plan but not a good short-term version; the Teleoperated Mobile Anti-Armor (TMAP) Program has both
4. Increase visibility of AI & robotics programs	4. Much progress made in past year
5. Automatic loader of ammunition in tanks	5. Generic Auto-Loader System is responsive to recommendation
6. Surveillance/sentry robot	6. TMAP development is well under way
7. Intelligent maintenance, diagnosis, and repair system	7. Hawk Missile Maintenance Tutor is responsive

2 ASSESSMENT OF INDIVIDUAL ARMY PROGRAMS

In recent years, the Army has used limited resources to develop selected robotics and artificial intelligence (AI) systems that would demonstrate performance in:

- improving soldier survivability,
- providing force multipliers,
- reducing operating and support costs, and
- enhancing training and education of personnel.

Army presentations to the committee indicated that the most important of the funded programs include:

- Teleoperated Mobile Anti-Armor Project (TMAP),
- Field Material Handling Robot (FMR),
- Soldier/Robot Interface Project (SRIP),
- Robotic Combat Vehicles, manned and unmanned (RCV),
- Advanced Ground Vehicle Technology (AGVT), and
- Hawk AI-based Maintenance Tutor (HAWK-MACH III).

In addition, the Army is supporting a small number of basic research projects in AI, done primarily by key university researchers, and a large number of expert system developmental projects, done primarily in Army laboratories and organizations.

While it was impossible to assess each of these programs in detail, the committee's consensus was that the projects had been well conceived and, if successfully completed, would significantly advance the introduction of AI and robotics into Army operations. A major concern of the committee, however, was that the planned present and future funding would not enable most of these programs (especially the robotic developments) to be translated into effective field-worthy systems in less than 10 to 15 years. Sufficient financial support could reduce the lead time to 5 years.

After hearing in October 1986 that large portions of the robotics programs were unfunded for FY 1987, the committee was pleased to learn that funds have been redirected to these programs. The Army's three primary robotics programs--the Teleoperated Mobile Anti-Armor Project, the Field Material Handling Robot, and Robotic Combat Vehicles--are now funded at levels at or near the amount requested. Those amounts are still modest, however, and more will be needed to develop the engineering prototypes and specifications that necessarily precede manufacturing.

The remainder of this chapter addresses each of the major programs, plus a worthwhile program of legged robots that the Army is not yet supporting. A summary of technical issues and a chart of the status of Army programs follows.

THE TELEOPERATED MOBILE ANTI-ARMOR PROGRAM

Members of the committee observed the concept evaluation program test of the Robotic Ranger conducted by the U.S. Army Infantry Board at Ft. Benning on June 11-12, 1986. The Ranger, developed by Grumman for the Missile Command, is the original demonstrator (two are now funded) of the TMAP concept. It is a teleoperated mobile vehicle equipped with weapon and reconnaissance systems.

The program is important not only because of the Ranger's potential as an effective anti-tank weapon, but also because the concept could be adapted for many other teleoperated mobile robot applications. Successful extensions of this technology could achieve large gains in soldier survivability at relatively low cost and offer the potential for significant force multiplication.

The value of developing and implementing large numbers of low-cost, expendable anti-armor (especially anti-tank) robots cannot be overestimated. The availability of such weapons to counterbalance the persistent numerical tank advantage of the Soviet bloc could have a profound effect on U.S. nuclear strategy. Moreover, the potential vulnerability of tanks to relatively inexpensive anti-tank robots could fundamentally affect Army tactics and weapons requirements for the field.

A major deficiency in the present TMAP program is the use of a fiber optic umbilical cord between the operator--

that is, the soldier--and the teleoperated vehicle. The fiber optic cord provides an excellent means for communications. It is wideband, does not require line of sight, does not radiate, and is not subject to interception. However, the cord has the serious problems of limited range and possible entanglement or cutting by the user or the enemy. This communication problem has received much attention in the past with little success. Nonetheless, because of the importance not only to the Army but to all the services, the committee believes that a sustained research and development effort should be undertaken immediately to develop a better solution.

Technologies made available in the past few years could be used in effecting secure communications without an umbilical cord. Specifically, the Army should investigate burst (i.e., non-continuous) communications, spread spectrum systems (which code and recapture signals in novel ways across wider bands), and other technologies for short-range secure communications. The broad scope and applicability of such research makes it an ideal candidate for support by the services and, in particular, DARPA. The Tech Base Enhancement for Autonomous Vehicles program, which was too new for the committee to receive a detailed briefing, plans to explore alternatives to to the fiber optic link.

In keeping with the strategy of combining short-term demonstrators with planned upgrades, it is appropriate that the Army plans to add a much higher degree of autonomy to the TMAP in the future. One promising opportunity would be to incorporate more advanced sensors for target acquisition, which would partially relieve the system of the need for continuous communication service with a human operator.

This program appears to be adequately funded to meet the stated near-term objectives. Full-scale engineering development, originally planned for early calendar year 1988, has--appropriately--been delayed until after a proof of principle demonstration in October 1988 and a 2 to 3 year preproduction engineering project beginning in 1989. Without intensive research to replace the umbilical cord, however, the expanded version of the TMAP is unlikely to be ready for deployment on schedule.

ROBOTIC MATERIAL HANDLING EQUIPMENT

The Army's projects in robotics material handling are the Field Material Handling Robot (FMR) and the Soldier/

Robot Interface Project (SRIP). Although the total amount of funding by the Army for FMR can be termed modest, at best, the establishment of individual projects on a joint basis with DARPA, plus exploitation of earlier NBS and industry IR&D efforts, presents an opportunity for successful transition to the 6.4 category of funding--engineering development with intent to produce for field use--by 1990.

Soldier/Robot Interface Project

A promising development in the Army's robotic material handling program is a very strong lightweight flexible robot arm with a deflection-compensating control system. The Army has gained leverage for its investment by involving Oak Ridge National Laboratories, with its experience in teleoperation in hostile environments, as a major participant. Odetics, Incorporated, is also a participant through the Small Business Innovation Research program.

Successful development would make possible many material handling robot operations in the field where low weight and high strength are of paramount importance; these attributes are currently missing from state-of-the-art industrial robots. The Army projects are exploring the many applications in which both small and very large versions of this principle would be most useful.

ROBOTIC COMBAT VEHICLES

The manned and unmanned Robotic Combat Vehicle (RCV) projects supported by DARPA, the Marines, and the Army are commendable. They are, however, all at a very early stage of development and will not be implemented until much later. For fully autonomous operation of the unmanned vehicle, the technical problems are severe. Automatic target acquisition, driving, and loading remain formidable problems, even for the manned tank with reduced manning. Scheduled funding indicates some opportunities for implementation, especially with the increase in FY 1987.

The Army is funding three RCV projects: Advanced Ground Vehicle Technology (AGVT), Tech Base Enhancement for Autonomous Machines (TEAM), and Robotic Command Center (RCC). The committee was briefed primarily on the AGVT; the other two are relatively new.

Advanced Ground Vehicle Technology

Committee members visited Martin Marietta in Denver on August 26, 1986. The trip involved both a visit to the Martin Marietta Autonomous Land Vehicle (ALV) laboratory and a DARPA-sponsored demonstration of a teleoperated Advanced Ground Vehicle built by FMC Corporation. The committee was unable to see another version of the Advanced Ground Vehicle Technology (AGVT), built by General Dynamics, that was demonstrated subsequently.

The committee was pleased to note the leverage that the AGVT gained by integrating the results of FMC Corporation's IR&D program with DARPA research on the ALV. The demonstration of the teleoperated vehicle was encouraging in that it will help set realistic requirements for robotic systems capable of performing combat and combat support missions.

The contrast between the terrain on which the ALV was operating and that on which it will eventually have to operate suggests that the ALV is a long-term solution that will not have useful short-term applications. The ALV was able to assist in developing the obstacle avoidance algorithms while operating on a flat surface with hay bales as obstacles; however, that is very different from the extremely rugged terrain that the ALV would eventually have to traverse to demonstrate its ability to operate autonomously. Despite the Army's strong need for autonomous vehicles, there are major problems ahead which are unlikely to be solved soon. As with most areas of research, the time to commercialize or achieve field use is longer than desired.

Although the committee understands that DARPA intends to devote its resources and efforts solely to autonomous mobility, teleoperation still appears to be the answer for the immediate future. That is, people will still be needed to operate the system for some time to come. That being the case, how do we make the operator (and the system) safer and more efficient? Specific needs are:

- more concern about counter-countermeasures: non-cable, non-line-of-sight, jam-resistant, highly reliable and secure data links with low probability of intercept;
- better stereoscopic or 3D vision to improve the teleoperator's ability to identify ditches, boulders, and other obstacles that could be obscured by vegetation or smoke; and

- better and easier map reading systems as aids to operators.

HAWK MAINTENANCE TUTOR

The development by the Army Research Institute (ARI) of the Hawk Maintenance Tutor is highly commendable. This system uses AI expert system technology for maintenance training. It can satisfy the need for training highly skilled technicians, who are in short supply now and for the foreseeable future. Quality of training and considerably reduced training time are impressive outcomes of this program. Further, this system can serve as a model for many other tutoring systems required for even more sophisticated high-technology weapon systems.

In addition to the strong applications-based uses of AI, such as for the Hawk Mach-III intelligent maintenance tutor, ARI is making good use of AI and expert systems in association with intelligent computer-aided instruction (ICAI) and is sponsoring a coherent Army program of research at some of the finest universities to help develop the needed training aids.

LEGGED MACHINES

The Army and DARPA have concentrated their support on the development of mobile vehicles that are either wheeled or tracked. DARPA is supporting several legged vehicle programs at Carnegie-Mellon and Ohio State Universities. These programs, however, are still in the research stage, far from use in practical situations. It appears that compact legged (or composite wheeled and legged) vehicles could have significant advantages in rough terrains.

About 5 years ago, Odetics Incorporated unveiled a teleoperated "functionoid"--a six-legged vehicle controlled by a person with computer assistance. This clever machine was creatively designed to be strong and versatile. However, it moved slowly and had no sensory feedback to help control it autonomously. Since then, the company has added a stabilizing sensory system to maintain equilibrium and leg sensors to enable the machine to maneuver over rough ground.

If a robot with multiple degrees of freedom (i.e., roll, pitch, and yaw, as well as movement along the x, y, and z axes) were added to the multi-legged platform, it could become a teleoperated aid to a soldier. Such a system could carry loads, perform repetitive tasks, be a stable mount for a sensor, rocket launcher, or other weapon, handle hazardous or toxic materials, and perform other tasks. The soldier-operator would control the machine with an appropriate human-engineered interface, either close by or remotely. The machine could have some degree of autonomy in repetitive tasks which the operator would train it to do. Such a system could be light (about 200 pounds) and strong (handle much more than its weight); it could be equipped with a small gasoline-driven electric recharger for a truck-sized battery to enable the system to operate over a good part of a day.

We recommend that the Army carefully assess the potential of such a legged, teleoperated "soldier's assistant." It could multiply the effectiveness of the armed forces--or reduce the number of soldiers needed--in the near future. Further, this work is an example of the advantages to the Army of monitoring technical developments in industry and adapting them to Army needs.

SUMMARY OF TECHNICAL AREAS

The review of selected Army programs and the site visits brought out seven technical areas that the Army should consider including in its robotics and AI programs:

1. Communication links for teleoperation,
2. Stereoscopic vision,
3. Map reading,
4. Sensors for target acquisition,
5. Deflection-compensated robot arms,
6. Stabilizing systems, and
7. Small, self-contained power sources.

The Army should carefully review the seven areas to answer these questions:

- Do programs in the Army or other service cover each area properly?
- Is more research and development needed?
- What other programs could use these technologies?

1. Communication Links for Teleoperation

Two of the Army's three major robotic programs--the Robotic Combat Vehicle and the Teleoperated Mobile Anti-Armor Program--need a broadband, jam-resistant, secure, and reliable data link. In particular, the TMAP's use of a fiber optic cable seriously limits the environments in which it can be used. The shielded cable by itself provides an excellent non-radiating, broad bandwidth (with three-dimensional capability) data link. However, optical cables and wires are not appropriate for field use. They have limited range and can easily be destroyed by either the enemy or the user. Increased research is needed to find durable and secure communication links.

2. Stereoscopic Vision

Teleoperation by scope observation and feedback showed the need for better three-dimensional or depth perception. This was confirmed by discussions with operators who have had many hours of training. Because teleoperation will be used for a long time, the Army should be working to improve stereoscopic vision.

3. Map Reading

At the sites, the soldiers' difficulties in reading were often mentioned as a fundamental problem. Proper map reading for land navigation is basic to teleoperation. Today's technologies should offer many ways to expedite and improve map reading, possibly with operator-assist systems and electronic storage and retrieval using optical disks. The ability to read maps "interactively" could be the technological solution to this continuing problem.

4. Sensors for Target Acquisition

Robotic sentries will be in positions where sensitive target sensors for object detection, identification, and location could add greatly to the robots' survival and ability to react quickly. The work needed in this area is to apply sensors--primarily imaging sensors--and to improve algorithms for dealing with noisy signals. Good sensors could give the robot a degree of autonomy without requiring constant communications to the teleoperator. They could also permit the robotic sentry to operate in nuclear, biological, or chemical (NBC) contamination or to warn of the presence of NBC contaminants.

5. Deflection-Compensated Robot Arms

Deflection compensation systems--including lasers, multi-linkages, and perhaps others--are being developed for robot arms. These devices could make commercial as well as Army-developed robot systems successful in areas where deflections of the robot arms would have caused some failures. People working in the application of robotics should understand these compensating principles.

6. Stabilizing System

Teleoperated vehicles will be exposed to terrain that challenges balance. It is important, therefore, to understand stabilizing systems. A stabilizing system has been developed for a multi-legged robot. This suggests that the principle should be studied for possible applications to teleoperated vehicles.

7. Small, Self-Contained Power Source

Teleoperated vehicles, especially small systems, have relatively short power life. Small, self-contained power sources could add greatly to the useful field time. A combination of intelligent sensors with teleoperation could give the best timing for recharging.

TABLE 2 Status of Selected Army Projects in Robotics and Artificial Intelligence, June 1987

PROGRAM	PROJECT	PROGRESS SINCE 1983	ORGANIZATIONAL AND TECHNICAL ISSUES
Teleoperated Mobile Anti-Armor Program	Robotic Ranger	Short-term demonstrator tested for battlefield; proof of principle planned October 1988	Optical cables not acceptable in field; need improved stereoscopic vision and advanced sensors for target acquisition
Robotic Material Handling	Field Material Handling Robot	Early development; transition to 6.4 by 1990 possible if work with DARPA & use of NBS work succeed	Total program funding should be increased
	Soldier/Robot Interface Project	Potential payoff of deflection-compensated flexible robotic arm	
Robotic Combat Vehicles	Advanced Ground Vehicle Technology	Demonstrator is functional; map reading system & terrain analysis under development; full autonomy still long-term possibility; teleoperation only option for near term	Not demonstrably prepared for very rugged terrain; need better 3D vision; tenuous communication links are vulnerable to interception
Hawk Maintenance Tutor	Hawk Mach-III Intelligent Maintenance Tutor	Army-sponsored university research is well planned and promising; strong applications of AI & expert systems	

3 THE ARMY ENVIRONMENT FOR ROBOTICS AND ARTIFICIAL INTELLIGENCE

The environment for research and development of robotics and artificial intelligence in the Army appears to have become much more robust since the committee's original study. Increased interest and activity in the application of robotics and artificial intelligence (AI) is evident at many levels of the Army.

This chapter addresses the management and organizational issues that apply across the specific robotics and artificial intelligence programs. These include inter- and intra-agency cooperation, the need for leadership and a champion, funding of applications, and industrial applications. The most urgent of the Army needs--education and training--is the subject of chapter 4.

INTER- AND INTRA-AGENCY COORDINATION

The committee commends the Army for its heightened awareness of AI and robotics programs. This awareness is reflected in the 1987 establishment of technology base groups for robotics and AI, the 1986 survey of Army AI projects compiled by the Army Research Office, the 1986 summer workshop, the accomplishments of the AI and robotics programs described in the preceding chapter, and enhanced training using AI techniques. The survey, however, tends to overstate the extent of the Army programs. The next workshop should broaden its scope with respect to invitees. Some of the best-known people in the field were not present at the first one, nor were their organizations represented. Nonetheless, the Army has shown a commendable initiative in seeking other sources of funding and assistance, especially in several cooperative efforts with DARPA and in exploitation of work either completed or under way at NBS, in universities, and in industry IR&D.

The TRADOC definition of AI and robotic requirements shows clear evidence of improvement. The designation of TRADOC "lead agents" will enhance the coordination of battlefield robotic development. Similar improvement is noted in the communication between the user and the materiel developers in the Army Materiel Command. Cooperation has improved markedly between the Army research community and DARPA, and between the Army and the other services.

As noted in chapter 2, the Army has joint projects not only with DARPA, but also with the Navy, Department of Energy, and the National Bureau of Standards, and a joint project is being discussed with the Marines. Although sponsored by DARPA, as a ground vehicle the Autonomous Land Vehicle program is of significant interest to the Army. Here government (DARPA, Army), industry (General Electric, SRI-International, Honeywell, Hughes, ADS, Martin Marietta), and academia (Carnegie-Mellon and Columbia Universities, Massachusetts Institute of Technology, and the Universities of Southern California, Rochester, Maryland, and Massachusetts) are cooperating in a move toward common robotic research and development goals. Concurrently, it promotes the educational environment and increases the involvement of people in AI and robotics as a discipline.

The committee noted an increasing division of the Army's R&D activity into robotics on the one hand and artificial intelligence on the other, with a third area that constitutes the intersection of the two fields. This is a natural and inevitable delineation of activity that can make all related activity more manageable, given top-level oversight and coordination.

Communications and networking need to be improved. Such a capability would provide the Army with a critical mass of expertise, geographically distributed. It could offer an intellectual environment that enabled individuals worldwide to work with colleagues in the same field, even when not located at the same laboratory. Networks such as MILNET and ARPANET, and electronic mailboxes, offer the opportunity for Army civilian and military personnel to get hands-on experience in this technology. The establishment of the Robotics and Artificial Intelligence Database is a step in this direction.

THE NEED FOR LEADERSHIP AND A CHAMPION

The demonstrator program recommended by the committee in its original report would be a necessary first step in the Army's exploitation of the tremendous potential that current technology offers. However, the committee finds that the Army's demonstrator program is funded at a level considerably lower than recommended in its original report. In fact, the current funding of this program appears to fall short of the critical mass of resource concentration required to demonstrate and exploit current technology before it becomes obsolete.

Only in FY 1987 did the Army appear to have moved toward centralizing responsibility for coordinating its robotics and AI programs. Two technology base groups--one for robotics, one for AI--were established at the Laboratory Command. The Army is to be commended for this increase in the visibility and coordination of these programs.

The preparation by TRADOC and AMC of an Army master plan for robotics provides an important tool for coordination. It will allow comprehensive management and direction in the application of robotics on the battlefield. Although the plan appears to be developing from the ground up rather than by edict from the top down, this type of creative work is to be encouraged. It should involve the assignment of a capable person to oversee, coordinate, evaluate, plan, manage, and promote Army robotic systems.

The demonstrator program needs strong leadership at the departmental level, both to integrate the Army's research and to justify the program to Congress. Vigorous, articulate leadership of the program at the Department of the Army level is essential to encourage Congress to supply the risk capital required to exploit new technology. The potential benefits of the near-term demonstrator program justify stronger central management and leadership of this program.

FUNDING OF APPLICATIONS

Engineering development--intention to produce full scale (6.4)--can be a very expensive phase of a program. If 6.4 funding is not carefully planned and given high-level support in advance, there is a danger that research and development (6.1, 6.2, 6.3, and 6.3a) will be reduced to obtain the 6.4 funds. If this happens, then the Army will be "eating its own seed corn" for the future. Investing in projects that integrate the most promising technology-based programs with user needs in the field seems to be the most prudent strategy at this time, provided that requirements documentation is started early and completed in time for early initiation of full-scale engineering development.

Another consideration, particularly with regard to large research or leapfrog areas, is that field-use experience will keep the research and development activities moving in the proper direction. However, this type of field testing must be done carefully. People other than the researchers (Army, Congress, and others) must be given enough background information to understand that the testing and experimentation is for knowledge of field use and not that a proven product is being tested.

INDUSTRIAL APPLICATIONS

It is interesting to note that the Japanese are currently engaged in a $100 million, 5-year program to develop prototypical robotic systems for service applications (not for factory automation). Construction, fire fighting, security, hazardous material handling, outdoor painting, and institutional cleaning are some of the applications envisaged. Although these are not simple applications, their requirements are considerably simpler than those for military service. Yet the funding level is far greater and more realistic than the Army's projections. It is thus very likely that many commercial Japanese robots will be performing functions in outdoor environments years before a similar robotic system is generally available in the U.S. Army.

In all advanced economies, private industry has clearly demonstrated the increased quality and productivity that can result from automation. Unlike tactical battlefield applications, which require enormous R&D expenditures by the Army to meet unique needs and operating conditions, the Army's industrial (internal manufacturing and material handling) applications resemble those of the private sector. Consequently, the time, money, and risk necessary to realize the expected benefits are relatively low. The committee was therefore surprised by the Army's apparently low usage of industrial robots. Several calls to various parts of the Army pointed to the Manufacturing and Methods Technology (MMT) program as the primary source of industrial robotic technology. This program has funded few robotic projects in the past and funds even fewer today. If other programs are not addressing this issue, we recommend increased use of the available technology within Army plants, depots, and arsenals.

In addition to industrial robots funded by the Army, robots developed independently by companies could be adapted for Army needs. The legged robots described in the previous chapter are one example; the FMC Corporation's research drawn on by the AGVT program is another. It is worthwhile for the Army to monitor such industrial developments.

Although increased use of industrial robots is warranted by the expected productivity improvements alone, the robots are also an important adjunct to the Army's education and training programs. Novices who have received formal AI or robotics training can successfully augment their understanding by implementing these applications--thus broadening the skill base within the Army--as they acquire the additional in-depth knowledge generally required for the more sophisticated tactical applications. Having completed a lower-risk project, this cadre of trained specialists can advance to other projects, leaving a spin-off group of technicians and operators behind, multiplying the training effect.

Implementing such a progression plan might include establishing one or more sites to demonstrate the solution to a generic, frequently occurring Army industrial problem. Serving first as a hands-on training aid, the site would also be a model for other installations throughout the Army. Careful selection of demonstration applications can also include the related technologies (vision, force sensing, etc.) which are the components of the corresponding tactical systems under development.

4 EDUCATION AND TRAINING

THE URGENT ARMY NEED

As science and technology grow in sophistication and as accelerating technological change becomes essential to the nation's defense, the Army will have correspondingly greater needs for education. It will require not only specialized education and training of personnel in the short term, but also the career-long education that must provide continuing career opportunities for growth.

Universities and colleges cannot meet these increased needs unaided. In the future, government and industry will need to assume more responsibility for maintaining current scientific and technical expertise--through research, development, education, and training. While this statement is more broadly applicable, it is particularly true of the Army's artificial intelligence (AI) and robotics programs.

After hearing the details of these programs, the committee concluded that the Army must significantly improve the quality and quantity of its educational and training programs. Personnel acquisition, retention, and development are a critical Army problem at present, and they will continue to pose problems unless significant changes occur. The Army does not offer competitive salaries for civilian or military personnel, it does not in general project the image of a high-technology environment, and it does not offer attractive career opportunities in robotics and AI. Further, the Army seems too dependent on short-term commitments by ROTC graduates, short courses, and contractor contacts for its expertise. A smart customer must be a smart doer, and the Army's internal skills and expertise must be increased.

In robotics and AI, the Army's efforts in education and research lag behind those of the Navy and Air Force not only in perception but also in fact. The long established Navy AI Center and the Northeast AI Consortium sponsored by the Air Force have no Army counterpart. Further, the Naval Postgraduate School, coupled with the numbers of service personnel sent for Master's and Doctoral degrees by the Air Force and Navy convey to civilian and military recruits the value placed on education.

UNIVERSITY CENTERS SPONSORED BY THE ARMY

The Army established Centers for AI at the University of Texas at Austin and the University of Pennsylvania in 1984. Although, in the opinion of the committee, these are not yet recognized as major research centers, they are operating satisfactorily. The productivity figures provided indicate normal professorial output in terms of publications and students. Greater effort is required, however, to transform each into a national Center for AI.

Our major concern is that the long-term funding of these Centers, by either the Army or the respective universities, is far from assured. Because the time required to earn a degree ranges from about 4 years for the undergraduate degree to as much as 8 years for the doctorate, funding cycles of such major Centers must be guaranteed for the long term. Otherwise, the programs should not be initiated, in fairness to the university, its faculty, and the students. From the Army's perspective, long-term financing is required because there is no return on short-term financing.

These centers need more publicity. Other centers, similar to those of the Army, have been and will continue to be funded. These include, among others, the Engineering Research Centers (ERCs) sponsored by the National Science Foundation (NSF) and the University Research Initiatives sponsored by the Department of Defense. In this competitive environment, the Army must do its part not only to sponsor but actively to promote its Centers in the hope of getting the most talented people.

ARMY INTERNAL EDUCATION PROGRAMS

An appropriate step taken by the Army is the establishment of an AI capability at West Point. The recent allocation of 6 to 8 research slots in this discipline is a major step forward. The committee suggests that these slots be used to teach (and thereby train Army officers in) AI as well as to do research.

In addition to this offering at West Point, the committee believes that such an opportunity should be made available at the Army's Command and General Staff College at Ft. Leavenworth, which now offers advanced degrees. Mid-career Army personnel in the mainstream of the Army's business attend the Staff College, and great benefit would be derived by starting an AI program there. Although West Point excels in military history and leadership programs, it is not chartered as a research institution. Consequently, it could not be expected to be a leader in AI research. Nevertheless, the coupling of West Point's program with the offering at Ft. Leavenworth would be a valuable addition to the Army's internal educational thrust.

The committee believes that Army career attractiveness and opportunities for civilians in AI and robotics are insufficient to acquire and retain key personnel. The Army must do more if it is to become a reservoir for talented people doing Army-related research and development in these fields. Educational programs must be long-term and include career-long education, rather than just 1-day or 2-week course offerings. Regular programs must be established for renewal of both military and civilians; they should include sabbaticals for advanced degrees or regular retraining. The committee noted that the Army has institutionalized certain areas. For example, every year a certain number of people are sent to Syracuse to be trained in comptrollership, while others go to Florida State for logistics management or to the University of Illinois for civil engineering. Perhaps it is time to institutionalize robotics and AI in the same way.

The absence of AI efforts from the University Research Initiative (URI) program is a missed opportunity. The URI program offers a unique opportunity for the services to concentrate large sums of long-term support in key technical areas of interest to them. The area of intelligent control, awarded to a consortium of MIT, Brown University, and Harvard, comes closest but is not mainstream AI. The existence of the two university Centers in AI does not constitute sufficient effort. If the FY 1986 URI program is followed by a second solicitation, as is being considered, the Army's need for further research in AI makes it an ideal candidate for a URI area.

Praiseworthy plans for a TRADOC-supervised, contractor-operated AI Applications Center sponsored by the Army were presented at the committee's January 1986 meeting. By the October 1986 meeting, however, plans for the Center had been dropped. (Note that this Applications Center differs from the AI Center at the Pentagon, about which the committee was not briefed.)

The transience of such proposals points to the fundamental issue--<u>the incompatibility of the Army's long-term planning and resources with educational time constants, complicated by the high rate of technological change</u>. The Army needs to recognize the importance of long-term support, identify key program areas, and stick with them; the uncertain status of the AI Applications Center suggests that the Army has not yet done this. As originally proposed, the Center was a good idea and we support it.

ARMY INTERNAL TRAINING PROGRAMS

The programs geared toward short-term training in AI and robotics generally seemed to be stronger than the Army's long-term educational activities. The Army recognizes that the technology is evolving and that, as a result, more intensive and frequent training of its personnel is required. Maintenance and support of equipment will become a major issue in the future, and the Army will need to upgrade its training even further.

The Army also has advanced to the use of technology such as CAI (computer-aided instruction) as teaching tools and training aids and is to be complimented on its progress. This is true not only of the Hawk Missile Maintenance Tutor described in the previous chapter, but also of the AI Center of Excellence, programs at Fort Gordon, and an entry level course at Fort Lee.

5 CONCLUSIONS AND RECOMMENDATIONS

CONCLUSIONS

1. Compared with 4 years ago, the Army has considerably improved its awareness and its implementation of AI and robotics development programs. These projects cover a restricted range of applications, including robot mobile vehicles, both teleoperated and autonomous; material handling robot systems; and a number of expert systems. Although many more applications that have been identified are not funded, the funded projects could be very useful for initiating and implementing future systems.

2. The Army has wisely concentrated its financial and managerial resources on a small number of projects. Although this is partly in agreement with our 1983 recommendation, the time period will be longer than the 2 to 3 years we had hoped for. Therefore, more funding will be needed to bring these projects to use in the field. We still believe that it is important to get early introduction of successful, simple robotic systems.

3. Robotics and AI still seem not to have visibility or champions at the highest levels in the Army. These technologies have tremendous potential to reduce hazards, increase individual effectiveness, reduce costs, and provide needed high-technology improvements in military and logistic operations. The establishment of AI and robotics technology base groups and the compilation of a master plan by TRADOC may reflect progress toward solving this problem. Without high-level support, the projects will not receive sufficient funding.

4. Education at all levels is the Army's greatest need as it adopts robotics and AI. With the growing, more integral role of the Armed Services in the nation's research and development effort, the Army needs to enhance its ability to attract and retain key personnel, both civilian and military. Career-long education and training is the key to a specialized Army work force that provides scientific and technological leadership, during peace and war.

5. The Army's work in robotics and artificial intelligence needs periodic external review and oversight. The Army Science Board is an obvious choice to be the focal point, but the task appears to require too much work for the board itself. The board, however, could pull together the people needed to perform the reviews.

6. Technical issues that the Army needs to address are:

- Communication links for teleoperation,
- Stereoscopic vision,
- Map reading,
- Sensors for target acquisition,
- Deflection-compensated robot arms,
- Stabilizing systems, and
- Small, self-contained power sources.

7. The Army has paid insufficient attention to utilizing existing robot technology for performing repetitive tasks in rear-echelon posts, depots, and warehouses. These tasks could include loading, unloading, and packaging. Mounting existing programmable robots on vehicles such as trucks, jeeps, or wheeled platforms would provide means of performing repetitive tasks with minimum training required if existing hand-held interface boxes that offer "training by doing" were used. Many tasks do not need customized robots that conform to very expensive specifications for battlefield survivability. Use of available technology would significantly decrease the time required to familiarize large numbers of soldiers with robots and their capabilities and programming and for the initiation of many new applications by the creative few.

8. Long-term planning and commitments are missing from Army AI and robotic educational programs. Evidence of this is that continued support for the AI Centers at the

University of Pennsylvania and the University of Texas is not certain, and a proposal for an Army AI center was dropped.

RECOMMENDATIONS

1. Increase the financial support for near-term robotic applications: Teleoperated Mobile Anti-Armor Project, Field Material Handling Robot, Soldier/Robotic Interface Prject. It appears possible to have field-useful systems within 5 years if sufficient funds are made available. In particular, concentrate attention on the TMAP project, with greatly increased emphasis on the development of counter-countermeasures and secure, survivable, non-jammable data links from the outset.

2. Establish a central, high-level Army organizational unit responsible for integrating and monitoring all AI and robotics R&D programs and guiding the subsequent procurement of fieldable systems.

3. Mount a serious effort to develop secure, survivable counter-countermeasures or secure, reliable, non-jammable, multidirectional communications (data links). These complex problems must be addressed from the outset--and solved--if the systems sought are to possess true operational utility.

4. Initiate studies aimed at the exploitation of existing legged mobile robot technology. The potential combination of deflection-compensating robot arms and automatic attitude stabilization in rough terrain could significantly advance TMAP application.

5. Initiate far more comprehensive educational and training programs for officers and technicians in AI and robotics, emphasizing the former.

6. Establish one or more university centers of excellence for AI and robotics, as part of the programs sponsored by the Army Research Office. Because these technologies are of interest to all the military services, co-sponsorship of such centers, or coordination under the University Research Initiative or other program, would be desirable.

7. Initiate a short study to determine the potential utility of equipping an existing Army truck or other vehicle with a commercially available, programmable robot for loading/unloading operations at depots or in the field.

LIST OF ACRONYMS

AGVT	Advanced Ground Vehicle Technology
AI	Artificial Intelligence
ALV	Autonomous Land Vehicle
AMC	Army Materiel Command
ARI	Army Research Institute
ARO	Army Research Office
CAI	Computer-Aided Instruction
DARPA	Defense Advanced Research Projects Agency
ERC	Engineering Research Center
FMR	Field Material Handling Robot
FY	Fiscal Year
ICAI	Intelligent Computer-Aided Instruction
IR&D	Independent Research & Development
MMT	Manufacturing Methods and Technology
NBC	Nuclear, Biological, and Chemical
NBS	National Bureau of Standards
NSF	National Science Foundation
R&D	Research and Development
RCC	Robotic Command Center
RCV	Robotic Combat Vehicles
SRIP	Soldier/Robot Interface Project
TEAM	Tech Base Enhancement for Autonomous Machines
TMAP	Teleoperated Mobile Anti-Armor Project
TRADOC	Training and Doctrine Command
URI	University Research Initiative